Higher
Mathematics

2002 Exam
Units 1, 2 and 3 Paper 1 (Non-calculator)
Units 1, 2 and 3 Paper 2

2002 Winter Diet Exam
Units 1, 2 and 3 Paper 1 (Non-calculator)
Units 1, 2 and 3 Paper 2

2003 Exam
Units 1, 2 and 3 Paper 1 (Non-calculator)
Units 1, 2 and 3 Paper 2

2004 Exam
Units 1, 2 and 3 Paper 1 (Non-calculator)
Units 1, 2 and 3 Paper 2

2005 Exam
Units 1, 2 and 3 Paper 1 (Non-calculator)
Units 1, 2 and 3 Paper 2

Leckie×Leckie

First exam published in 2002.

Published by Leckie & Leckie, 8 Whitehill Terrace, St. Andrews, Scotland KY16 8RN tel: 01334 475656 fax: 01334 477392 enquiries@leckieandleckie.co.uk www.leckieandleckie.co.uk

ISBN 1-84372-347-6

A CIP Catalogue record for this book is available from the British Library.

Printed in Scotland by Scotprint.

Leckie & Leckie is a division of Granada Learning Limited, part of ITV plc.

Acknowledgements

Leckie & Leckie is grateful to the copyright holders, as credited at the back of the book, for permission to use their material.
Every effort has been made to trace the copyright holders and to obtain their permission for the use of copyright material.
Leckie & Leckie will gladly receive information enabling them to rectify any error or omission in subsequent editions.

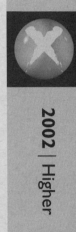

2002 | Higher

[BLANK PAGE]

X100/301

NATIONAL QUALIFICATIONS 2002	MONDAY, 27 MAY 9.00 AM – 10.10 AM	**MATHEMATICS HIGHER** Units 1, 2 and 3 Paper 1 (Non-calculator)

Read Carefully

1   **Calculators may <u>NOT</u> be used in this paper.**

2   Full credit will be given only where the solution contains appropriate working.

3   Answers obtained by readings from scale drawings will not receive any credit.

SCOTTISH QUALIFICATIONS AUTHORITY

FORMULAE LIST

Circle:

The equation $x^2 + y^2 + 2gx + 2fy + c = 0$ represents a circle centre $(-g, -f)$ and radius $\sqrt{g^2 + f^2 - c}$.

The equation $(x - a)^2 + (y - b)^2 = r^2$ represents a circle centre (a, b) and radius r.

Scalar Product: $a.b = |a|\,|b| \cos \theta$, where θ is the angle between a and b

or $a.b = a_1b_1 + a_2b_2 + a_3b_3$ where $a = \begin{pmatrix} a_1 \\ a_2 \\ a_3 \end{pmatrix}$ and $b = \begin{pmatrix} b_1 \\ b_2 \\ b_3 \end{pmatrix}$.

Trigonometric formulae:

$$\sin (A \pm B) = \sin A \cos B \pm \cos A \sin B$$
$$\cos (A \pm B) = \cos A \cos B \mp \sin A \sin B$$
$$\sin 2A = 2\sin A \cos A$$
$$\cos 2A = \cos^2 A - \sin^2 A$$
$$= 2\cos^2 A - 1$$
$$= 1 - 2\sin^2 A$$

Table of standard derivatives:

$f(x)$	$f'(x)$
$\sin ax$	$a\cos ax$
$\cos ax$	$-a\sin ax$

Table of standard integrals:

$f(x)$	$\int f(x)\, dx$
$\sin ax$	$-\dfrac{1}{a}\cos ax + C$
$\cos ax$	$\dfrac{1}{a}\sin ax + C$

ALL questions should be attempted.

Marks

1. The point P(2, 3) lies on the circle $(x + 1)^2 + (y - 1)^2 = 13$. Find the equation of the tangent at P.

   4

2. The point Q divides the line joining P(−1, −1, 0) to R(5, 2, −3) in the ratio 2 : 1. Find the coordinates of Q.

   3

3. Functions f and g are defined on suitable domains by $f(x) = \sin(x°)$ and $g(x) = 2x$.

   (a) Find expressions for:

     (i) $f(g(x))$;

     (ii) $g(f(x))$.

   2

   (b) Solve $2f(g(x)) = g(f(x))$ for $0 \le x \le 360$.

   5

4. Find the coordinates of the point on the curve $y = 2x^2 - 7x + 10$ where the tangent to the curve makes an angle of 45° with the positive direction of the x-axis.

   4

5. In triangle ABC, show that the exact value of $\sin(a + b)$ is $\dfrac{2}{\sqrt{5}}$.

   4

6. The graph of a function f intersects the x-axis at $(-a, 0)$ and $(e, 0)$ as shown.

   There is a point of inflexion at $(0, b)$ and a maximum turning point at (c, d).

   Sketch the graph of the derived function f'.

   3

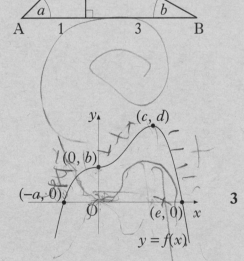

[Turn over for Questions 7 to 11 on *Page four*

Marks

7. (a) Express $f(x) = x^2 - 4x + 5$ in the form $f(x) = (x - a)^2 + b$.   2

   (b) On the same diagram sketch:
      (i)   the graph of $y = f(x)$;
      (ii) the graph of $y = 10 - f(x)$.   4

   (c) Find the range of values of x for which $10 - f(x)$ is positive.   1

8. The diagram shows the graph of a cosine function from 0 to π.

   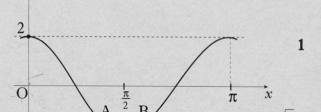

   (a) State the equation of the graph.   1

   (b) The line with equation $y = -\sqrt{3}$ intersects this graph at points A and B.

   Find the coordinates of B.   3

9. (a) Write $\sin(x) - \cos(x)$ in the form $k\sin(x - a)$ stating the values of k and a where $k > 0$ and $0 \le a \le 2\pi$.   4

   (b) Sketch the graph of $y = \sin(x) - \cos(x)$ for $0 \le x \le 2\pi$, showing clearly the graph's maximum and minimum values and where it cuts the x-axis and the y-axis.   3

10. (a) Find the derivative of the function $f(x) = (8 - x^3)^{\frac{1}{2}}$, $x < 2$.   2

    (b) Hence write down $\displaystyle\int \frac{x^2}{(8 - x^3)^{\frac{1}{2}}} \, dx$.   1

11. The graph illustrates the law $y = kx^n$.
    If the straight line passes through A(0·5, 0) and B(0, 1), find the values of k and n.   4

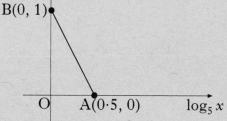

[END OF QUESTION PAPER]

X100/303

NATIONAL
QUALIFICATIONS
2002

MONDAY, 27 MAY
10.30 AM – 12.00 NOON

MATHEMATICS
HIGHER
Units 1, 2 and 3
Paper 2

Read Carefully

1 **Calculators may be used in this paper.**

2 Full credit will be given only where the solution contains appropriate working.

3 Answers obtained by readings from scale drawings will not receive any credit.

SCOTTISH
QUALIFICATIONS
AUTHORITY

FORMULAE LIST

Circle:

The equation $x^2 + y^2 + 2gx + 2fy + c = 0$ represents a circle centre $(-g, -f)$ and radius $\sqrt{g^2 + f^2 - c}$.

The equation $(x - a)^2 + (y - b)^2 = r^2$ represents a circle centre (a, b) and radius r.

Scalar Product: $\boldsymbol{a.b} = |\boldsymbol{a}|\,|\boldsymbol{b}| \cos\theta$, where θ is the angle between $\boldsymbol{a}$ and $\boldsymbol{b}$

or $\boldsymbol{a.b} = a_1 b_1 + a_2 b_2 + a_3 b_3$ where $\boldsymbol{a} = \begin{pmatrix} a_1 \\ a_2 \\ a_3 \end{pmatrix}$ and $\boldsymbol{b} = \begin{pmatrix} b_1 \\ b_2 \\ b_3 \end{pmatrix}$.

Trigonometric formulae:
$$\sin (A \pm B) = \sin A \cos B \pm \cos A \sin B$$
$$\cos (A \pm B) = \cos A \cos B \mp \sin A \sin B$$
$$\sin 2A = 2\sin A \cos A$$
$$\cos 2A = \cos^2 A - \sin^2 A$$
$$= 2\cos^2 A - 1$$
$$= 1 - 2\sin^2 A$$

Table of standard derivatives:

$f(x)$	$f'(x)$
$\sin ax$	$a\cos ax$
$\cos ax$	$-a\sin ax$

Table of standard integrals:

$f(x)$	$\int f(x)\,dx$
$\sin ax$	$-\dfrac{1}{a}\cos ax + C$
$\cos ax$	$\dfrac{1}{a}\sin ax + C$

ALL questions should be attempted.

Marks

1. Triangle ABC has vertices A(–1, 6), B(–3, –2) and C(5, 2).

   Find

   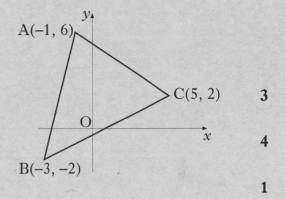

   (a) the equation of the line p, the median from C of triangle ABC.   **3**

   (b) the equation of the line q, the perpendicular bisector of BC.   **4**

   (c) the coordinates of the point of intersection of the lines p and q.   **1**

2. The diagram shows a square-based pyramid of height 8 units.

   Square OABC has a side length of 6 units.

   The coordinates of A and D are (6, 0, 0) and (3, 3, 8).

   C lies on the y-axis.

   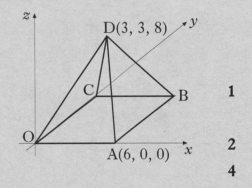

   (a) Write down the coordinates of B.   **1**

   (b) Determine the components of $\overrightarrow{DA}$ and $\overrightarrow{DB}$.   **2**

   (c) Calculate the size of angle ADB.   **4**

3. The diagram shows part of the graph of the curve with equation $y = 2x^3 - 7x^2 + 4x + 4$.

   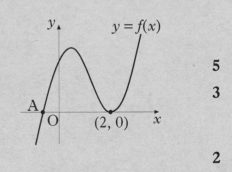

   (a) Find the x-coordinate of the maximum turning point.   **5**

   (b) Factorise $2x^3 - 7x^2 + 4x + 4$.   **3**

   (c) State the coordinates of the point A and hence find the values of x for which $2x^3 - 7x^2 + 4x + 4 < 0$.   **2**

   [Turn over

Marks

4. A man decides to plant a number of fast-growing trees as a boundary between his property and the property of his next door neighbour. He has been warned, however, by the local garden centre that, during any year, the trees are expected to increase in height by 0·5 metres. In response to this warning he decides to trim 20% off the height of the trees at the start of any year.

   (a) If he adopts the "20% pruning policy", to what height will he expect the trees to grow in the long run?   **3**

   (b) His neighbour is concerned that the trees are growing at an alarming rate and wants assurances that the trees will grow no taller than 2 metres. What is the minimum percentage that the trees will need to be trimmed each year so as to meet this condition?   **3**

5. Calculate the shaded area enclosed between the parabolas with equations $y = 1 + 10x - 2x^2$ and $y = 1 + 5x - x^2$.   **6**

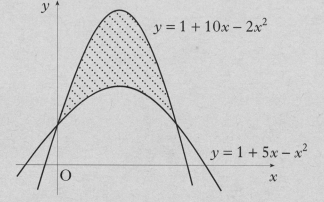

6. Find the equation of the tangent to the curve $y = 2\sin\left(x - \dfrac{\pi}{6}\right)$ at the point where $x = \dfrac{\pi}{3}$.   **4**

7. Find the x-coordinate of the point where the graph of the curve with equation $y = \log_3(x - 2) + 1$ intersects the x-axis.   **3**

8. A point moves in a straight line such that its acceleration a is given by $a = 2(4 - t)^{\frac{1}{2}}$, $0 \le t \le 4$. If it starts at rest, find an expression for the velocity v where $a = \dfrac{dv}{dt}$.   **4**

9. Show that the equation $(1 - 2k)x^2 - 5kx - 2k = 0$ has real roots for all integer values of k.   **5**

Marks

10. The shaded rectangle on this map represents the planned extension to the village hall. It is hoped to provide the largest possible area for the extension.

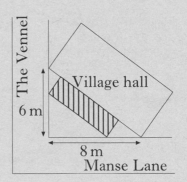

The coordinate diagram represents the right angled triangle of ground behind the hall. The extension has length l metres and breadth b metres, as shown. One corner of the extension is at the point $(a, 0)$.

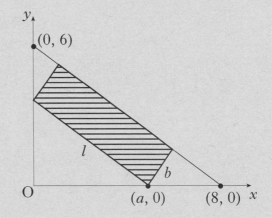

(a) (i) Show that $l = \frac{5}{4}a$.

   (ii) Express b in terms of a and hence deduce that the area, $A\,\text{m}^2$, of the extension is given by $A = \frac{3}{4}a(8 - a)$. **3**

(b) Find the value of a which produces the largest area of the extension. **4**

[END OF QUESTION PAPER]

[BLANK PAGE]

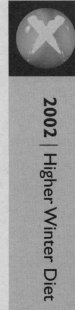

[BLANK PAGE]

W100/301

NATIONAL QUALIFICATIONS 2002

FRIDAY, 18 JANUARY 9.00 AM – 10.10 AM

MATHEMATICS HIGHER Units 1, 2 and 3 Paper 1 (Non-calculator)

Read Carefully

1 Calculators may **NOT** be used in this paper.

2 Full credit will be given only where the solution contains appropriate working.

3 Answers obtained by readings from scale drawings will not receive any credit.

SCOTTISH QUALIFICATIONS AUTHORITY

FORMULAE LIST

Circle:

The equation $x^2 + y^2 + 2gx + 2fy + c = 0$ represents a circle centre $(-g, -f)$ and radius $\sqrt{g^2 + f^2 - c}$.

The equation $(x - a)^2 + (y - b)^2 = r^2$ represents a circle centre (a, b) and radius r.

Scalar Product: $\quad a.b = |a|\,|b|\cos\theta$, where θ is the angle between a and b

$$\text{or} \quad a.b = a_1b_1 + a_2b_2 + a_3b_3 \text{ where } a = \begin{pmatrix} a_1 \\ a_2 \\ a_3 \end{pmatrix} \text{ and } b = \begin{pmatrix} b_1 \\ b_2 \\ b_3 \end{pmatrix}.$$

Trigonometric formulae:

$$\sin(A \pm B) = \sin A \cos B \pm \cos A \sin B$$
$$\cos(A \pm B) = \cos A \cos B \mp \sin A \sin B$$
$$\sin 2A = 2\sin A \cos A$$
$$\cos 2A = \cos^2 A - \sin^2 A$$
$$= 2\cos^2 A - 1$$
$$= 1 - 2\sin^2 A$$

Table of standard derivatives:

$f(x)$	$f'(x)$
$\sin ax$	$a\cos ax$
$\cos ax$	$-a\sin ax$

Table of standard integrals:

$f(x)$	$\int f(x)\,dx$
$\sin ax$	$-\dfrac{1}{a}\cos ax + C$
$\cos ax$	$\dfrac{1}{a}\sin ax + C$

ALL questions should be attempted.

Marks

1. (a) Find the equation of the straight line through the points A(−1, 5) and B(3, 1). **2**

   (b) Find the size of the angle which AB makes with the positive direction of the x-axis. **2**

2. (a) If $u = \begin{pmatrix} 1 \\ 7 \\ -2 \end{pmatrix}$ and $v = \begin{pmatrix} 1 \\ -2 \\ 1 \end{pmatrix}$, write down the components of $u + 3v$ and $u − 3v$. **2**

   (b) Hence, or otherwise, show that $u + 3v$ and $u − 3v$ are perpendicular. **2**

3. Find the equation of the tangent to the curve with equation $y = \frac{3}{x}$ at the point P where $x = 1$. **5**

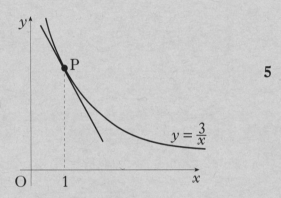

4. (a) Write down the exact values of $\sin\left(\frac{\pi}{3}\right)$ and $\cos\left(\frac{\pi}{3}\right)$. **1**

   (b) If $\tan x = 4\sin\left(\frac{\pi}{3}\right)\cos\left(\frac{\pi}{3}\right)$, find the exact values of x for $0 \le x \le 2\pi$. **2**

5. Given that $(x − 2)$ and $(x + 3)$ are factors of $f(x)$ where $f(x) = 3x^3 + 2x^2 + cx + d$, find the values of c and d. **5**

[Turn over

Marks

6. The side view of part of a roller coaster ride is shown by the path PQRS.
The curve PQ is an arc of the circle with equation $x^2 + y^2 + 4x - 10y + 9 = 0$.
The curve QRS is part of the parabola with equation $y = -x^2 + 6x - 5$.
The point Q has coordinates (2, 3).

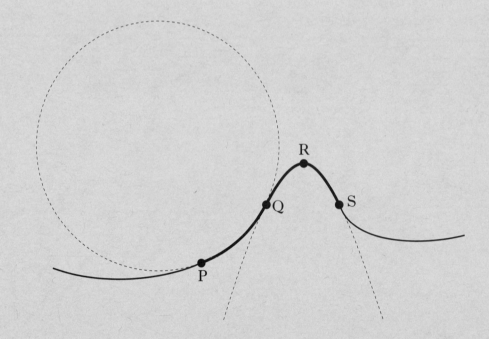

(a) Find the equation of the tangent to the circle at Q.   4

(b) Show that this tangent to the circle at Q is also the tangent to the parabola at Q.   2

7. Find $\displaystyle\int\left(\sqrt[3]{x} - \frac{1}{\sqrt{x}}\right)dx.$ 4

8. The diagram shows part of the graph of $y = 2^x$.

(a) Sketch the graph of $y = 2^{-x} - 8$.   2

(b) Find the coordinates of the points where it crosses the x and y axes.   2

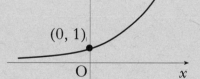

Marks

9. The function f, defined on a suitable domain, is given by $f(x) = \dfrac{3}{x+1}$.

    (a) Find an expression for $h(x)$ where $h(x) = f(f(x))$, giving your answer as a fraction in its simplest form.   **3**

    (b) Describe any restriction on the domain of h.   **1**

10. A function f is defined by $f(x) = 2x + 3 + \dfrac{18}{x-4}$, $x \neq 4$.

    Find the values of x for which the function is increasing.   **5**

11. PQRSTU is a regular hexagon of side 2 units. $\overrightarrow{PQ}$, $\overrightarrow{QR}$ and $\overrightarrow{RS}$ represent vectors **a**, **b** and **c** respectively.

    Find the value of **a**.(**b** + **c**).   **3**

    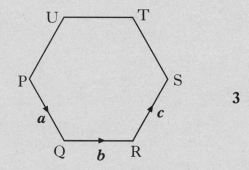

12. If $\log_a p = \cos^2 x$ and $\log_a r = \sin^2 x$, show that $pr = a$.   **3**

[END OF QUESTION PAPER]

[BLANK PAGE]

W100/303

NATIONAL
QUALIFICATIONS
2002

FRIDAY, 18 JANUARY
10.30 AM – 12.00 NOON

MATHEMATICS
HIGHER
Units 1, 2 and 3
Paper 2

Read Carefully

1 **Calculators may be used in this paper.**

2 Full credit will be given only where the solution contains appropriate working.

3 Answers obtained by readings from scale drawings will not receive any credit.

SCOTTISH
QUALIFICATIONS
AUTHORITY

FORMULAE LIST

Circle:

The equation $x^2 + y^2 + 2gx + 2fy + c = 0$ represents a circle centre $(-g, -f)$ and radius $\sqrt{g^2 + f^2 - c}$.

The equation $(x - a)^2 + (y - b)^2 = r^2$ represents a circle centre (a, b) and radius r.

Scalar Product: $a.b = |a|\,|b| \cos \theta$, where θ is the angle between a and b

or $a.b = a_1b_1 + a_2b_2 + a_3b_3$ where $a = \begin{pmatrix} a_1 \\ a_2 \\ a_3 \end{pmatrix}$ and $b = \begin{pmatrix} b_1 \\ b_2 \\ b_3 \end{pmatrix}$.

Trigonometric formulae:
$$\sin(A \pm B) = \sin A \cos B \pm \cos A \sin B$$
$$\cos(A \pm B) = \cos A \cos B \mp \sin A \sin B$$
$$\sin 2A = 2\sin A \cos A$$
$$\cos 2A = \cos^2 A - \sin^2 A$$
$$= 2\cos^2 A - 1$$
$$= 1 - 2\sin^2 A$$

Table of standard derivatives:

$f(x)$	$f'(x)$
$\sin ax$	$a\cos ax$
$\cos ax$	$-a\sin ax$

Table of standard integrals:

$f(x)$	$\int f(x)\,dx$
$\sin ax$	$-\dfrac{1}{a}\cos ax + C$
$\cos ax$	$\dfrac{1}{a}\sin ax + C$

ALL questions should be attempted.

Marks

1. The diagram shows a rhombus PQRS with its diagonals PR and QS.

    PR has equation $y = 2x - 2$.

    Q has coordinates $(-2, 4)$.

    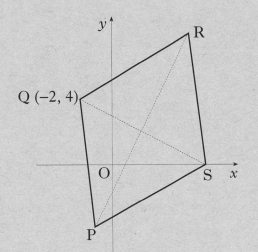

    (a) (i) Find the equation of the diagonal QS.

    (ii) Find the coordinates of T, the point of intersection of PR and QS.

    6

    (b) R is the point $(5, 8)$. Write down the coordinates of P.

    2

2. With reference to a suitable set of coordinate axes, A, B and C are the points $(-8, 10, 20)$, $(-2, 1, 8)$ and $(0, -2, 4)$ respectively.

    Show that A, B and C are collinear and find the ratio AB : BC.

    4

3. (a) Calculate the limit as n $\rightarrow \infty$ of the sequence defined by $u_{n+1} = 0 \cdot 9 u_n + 10$, $u_0 = 1$.

    3

    (b) Determine the least value of n for which u_n is greater than half of this limit and the corresponding value of u_n.

    2

4. (a) Write $\sqrt{3} \sin x° + \cos x°$ in the form $k \sin(x + a)°$ where $k > 0$ and $0 \leq a < 360$.

    4

    (b) Hence find the maximum value of $5 + \sqrt{3} \sin x° + \cos x°$ and determine the corresponding value of x in the interval $0 \leq x \leq 360$.

    2

5. Solve the equation $\cos 2x - 2 \sin^2 x = 0$ in the interval $0 \leq x < 2\pi$.

    4

    [Turn over

Marks

6. The graph of $f(x) = 2x^3 - 5x^2 - 3x + 1$ has been sketched in the diagram shown.

   Find the value of a correct to one decimal place.

   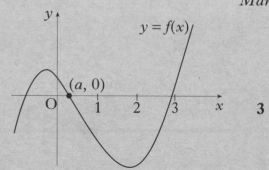

   3

7. A rectangular beam is to be cut from a cylindrical log of diameter 20 cm.

   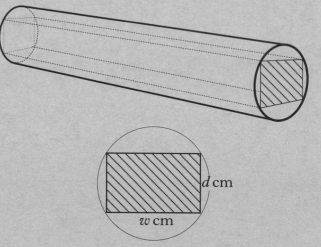

   The diagram shows a cross-section of the log and beam where the beam has a breadth of w cm and a depth of d cm.

   The strength S of the beam is given by

   $S = 1{\cdot}7\,w\,(400 - w^2)$.

   Find the dimensions of the beam for maximum strength.

   5

8. Find $\displaystyle\int_0^1 \left(\cos(3x) - \sin\left(\tfrac{1}{3}x + 1\right)\right) dx$ correct to 3 decimal places.

   3

9. A researcher modelled the size N of a colony of bacteria t hours after the beginning of her observations by $N(t) = 950 \times (2{\cdot}6)^{0{\cdot}2t}$.

   (a) What was the size of the colony when observations began?

   1

   (b) How long does it take for the size of the colony to be multiplied by 10?

   4

10. The line $y + 2x = k$, $k > 0$, is a tangent to the circle $x^2 + y^2 - 2x - 4 = 0$.

   (a) Find the value of k.

   7

   (b) Deduce the coordinates of the point of contact.

   2

Marks

11. An energy efficient building is designed with solar cells covering the whole of its south facing roof. The energy generated by the solar cells is directly proportional to the area, in square units, of the solar roof.

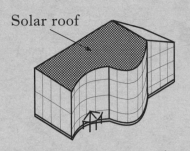

Solar roof

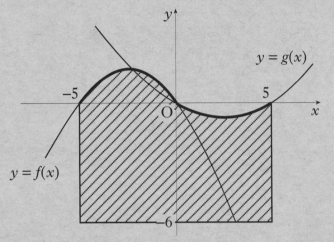

The shape of the solar roof can be represented on the coordinate plane as the shaded area bounded by the functions $f(x) = \frac{1}{4}\left(-x^2 - 5x\right)$, $g(x) = \frac{1}{12}\left(x^2 - 5x\right)$ and the lines $x = -5$, $x = 5$ and $y = -6$.

(a) Find the area of the solar roof.　　　　　　　　　　　　　　　　　7

(b) Ten square units of solar cells generate a maximum of 1 kilowatt.

What is the maximum energy the solar roof can generate in kilowatts (to the nearest kilowatt)?　　　　1

[END OF QUESTION PAPER]

[BLANK PAGE]

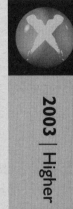

[BLANK PAGE]

X100/301

NATIONAL
QUALIFICATIONS
2003

WEDNESDAY, 21 MAY
9.00 AM – 10.10 AM

MATHEMATICS
HIGHER
Units 1, 2 and 3
Paper 1
(Non-calculator)

Read Carefully

1 **Calculators may <u>NOT</u> be used in this paper.**

2 Full credit will be given only where the solution contains appropriate working.

3 Answers obtained by readings from scale drawings will not receive any credit.

SCOTTISH
QUALIFICATIONS
AUTHORITY

FORMULAE LIST

Circle:

The equation $x^2 + y^2 + 2gx + 2fy + c = 0$ represents a circle centre $(-g, -f)$ and radius $\sqrt{g^2 + f^2 - c}$.

The equation $(x - a)^2 + (y - b)^2 = r^2$ represents a circle centre (a, b) and radius r.

Scalar Product: $a.b = |a|\,|b| \cos \theta$, where θ is the angle between a and b

or $a.b = a_1b_1 + a_2b_2 + a_3b_3$ where $a = \begin{pmatrix} a_1 \\ a_2 \\ a_3 \end{pmatrix}$ and $b = \begin{pmatrix} b_1 \\ b_2 \\ b_3 \end{pmatrix}$.

Trigonometric formulae:

$$\sin (A \pm B) = \sin A \cos B \pm \cos A \sin B$$
$$\cos (A \pm B) = \cos A \cos B \mp \sin A \sin B$$
$$\sin 2A = 2\sin A \cos A$$
$$\cos 2A = \cos^2 A - \sin^2 A$$
$$= 2\cos^2 A - 1$$
$$= 1 - 2\sin^2 A$$

Table of standard derivatives:

$f(x)$	$f'(x)$
$\sin ax$	$a\cos ax$
$\cos ax$	$-a\sin ax$

Table of standard integrals:

$f(x)$	$\int f(x)\,dx$
$\sin ax$	$-\frac{1}{a}\cos ax + C$
$\cos ax$	$\frac{1}{a}\sin ax + C$

ALL questions should be attempted.

Marks

1. Find the equation of the line which passes through the point (−1, 3) and is perpendicular to the line with equation $4x + y - 1 = 0$.

   3

2. (a) Write $f(x) = x^2 + 6x + 11$ in the form $(x + a)^2 + b$.

   2

   (b) Hence or otherwise sketch the graph of $y = f(x)$.

   2

3. Vectors $\boldsymbol{u}$ and $\boldsymbol{v}$ are defined by $\boldsymbol{u} = 3\boldsymbol{i} + 2\boldsymbol{j}$ and $\boldsymbol{v} = 2\boldsymbol{i} - 3\boldsymbol{j} + 4\boldsymbol{k}$.

   Determine whether or not $\boldsymbol{u}$ and $\boldsymbol{v}$ are perpendicular to each other.

   2

4. A recurrence relation is defined by $u_{n+1} = pu_n + q$, where $-1 < p < 1$ and $u_0 = 12$.

   (a) If $u_1 = 15$ and $u_2 = 16$, find the values of p and q.

   2

   (b) Find the limit of this recurrence relation as $n \to \infty$.

   2

5. Given that $f(x) = \sqrt{x} + \dfrac{2}{x^2}$, find $f'(4)$.

   5

6. A and B are the points (−1, −3, 2) and (2, −1, 1) respectively.

   B and C are the points of trisection of AD, that is AB = BC = CD.

   Find the coordinates of D.

   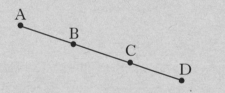

   3

7. Show that the line with equation $y = 2x + 1$ does not intersect the parabola with equation $y = x^2 + 3x + 4$.

   5

8. Find $\displaystyle\int_{0}^{1} \dfrac{dx}{(3x + 1)^{\frac{1}{2}}}$.

   4

9. Functions $f(x) = \dfrac{1}{x - 4}$ and $g(x) = 2x + 3$ are defined on suitable domains.

   (a) Find an expression for $h(x)$ where $h(x) = f(g(x))$.

   2

   (b) Write down any restriction on the domain of h.

   1

[Turn over for Questions 10 to 12 on *Page four*

Marks

10. A is the point (8, 4). The line OA is inclined at an angle p radians to the x-axis.

    (a) Find the exact values of:
       (i) $\sin(2p)$;
       (ii) $\cos(2p)$.

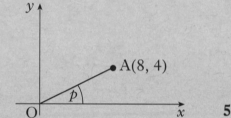

5

The line OB is inclined at an angle $2p$ radians to the x-axis.

    (b) Write down the exact value of the gradient of OB.

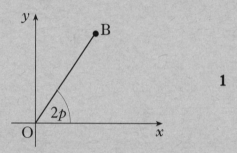

1

11. • O, A and B are the centres of the three circles shown in the diagram below.
    • The two outer circles are congruent and each touches the smallest circle.
    • Circle centre A has equation $(x - 12)^2 + (y + 5)^2 = 25$.
    • The three centres lie on a parabola whose axis of symmetry is shown by the broken line through A.

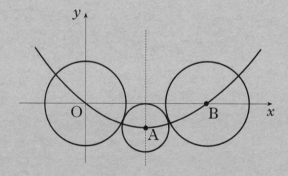

    (a) (i) State the coordinates of A and find the length of the line OA.   2
        (ii) Hence find the equation of the circle with centre B.   3

    (b) The equation of the parabola can be written in the form $y = px(x + q)$. Find the values of p and q.   2

12. Simplify $3 \log_e(2e) - 2 \log_e(3e)$ expressing your answer in the form $A + \log_e B - \log_e C$ where A, B and C are whole numbers.   4

[END OF QUESTION PAPER]

X100/303

NATIONAL
QUALIFICATIONS
2003

WEDNESDAY, 21 MAY
10.30 AM – 12.00 NOON

MATHEMATICS
HIGHER
Units 1, 2 and 3
Paper 2

Read Carefully

1   **Calculators may be used in this paper.**

2   Full credit will be given only where the solution contains appropriate working.

3   Answers obtained by readings from scale drawings will not receive any credit.

SCOTTISH
QUALIFICATIONS
AUTHORITY

FORMULAE LIST

Circle:

The equation $x^2 + y^2 + 2gx + 2fy + c = 0$ represents a circle centre $(-g, -f)$ and radius $\sqrt{g^2 + f^2 - c}$.

The equation $(x - a)^2 + (y - b)^2 = r^2$ represents a circle centre (a, b) and radius r.

Scalar Product: $a.b = |a|\,|b|\cos\theta$, where θ is the angle between a and b

or $a.b = a_1 b_1 + a_2 b_2 + a_3 b_3$ where $a = \begin{pmatrix} a_1 \\ a_2 \\ a_3 \end{pmatrix}$ and $b = \begin{pmatrix} b_1 \\ b_2 \\ b_3 \end{pmatrix}$.

Trigonometric formulae:

$$\sin (A \pm B) = \sin A \cos B \pm \cos A \sin B$$
$$\cos (A \pm B) = \cos A \cos B \mp \sin A \sin B$$
$$\sin 2A = 2\sin A \cos A$$
$$\cos 2A = \cos^2 A - \sin^2 A$$
$$= 2\cos^2 A - 1$$
$$= 1 - 2\sin^2 A$$

Table of standard derivatives:

$f(x)$	$f'(x)$
$\sin ax$	$a\cos ax$
$\cos ax$	$-a\sin ax$

Table of standard integrals:

$f(x)$	$\int f(x)\,dx$
$\sin ax$	$-\frac{1}{a}\cos ax + C$
$\cos ax$	$\frac{1}{a}\sin ax + C$

ALL questions should be attempted.

Marks

1. $f(x) = 6x^3 - 5x^2 - 17x + 6$.

   (*a*) Show that $(x - 2)$ is a factor of $f(x)$.

   (*b*) Express $f(x)$ in its fully factorised form.   4

2. The diagram shows a sketch of part of the graph of a trigonometric function whose equation is of the form $y = a \sin(bx) + c$.

   Determine the values of a, b and c.   3

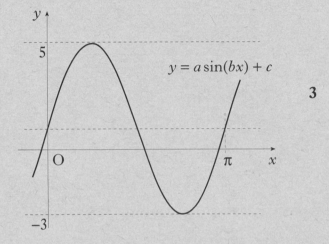

3. The incomplete graphs of $f(x) = x^2 + 2x$ and $g(x) = x^3 - x^2 - 6x$ are shown in the diagram. The graphs intersect at A(4, 24) and the origin.

   Find the shaded area enclosed between the curves.   5

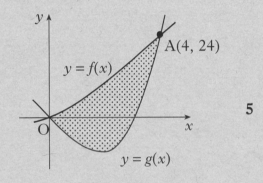

4. (*a*) Find the equation of the tangent to the curve with equation $y = x^3 + 2x^2 - 3x + 2$ at the point where $x = 1$.   5

   (*b*) Show that this line is also a tangent to the circle with equation $x^2 + y^2 - 12x - 10y + 44 = 0$ and state the coordinates of the point of contact.   6

[Turn over

Marks

5. The diagram shows the graph of a function *f*.

   f has a minimum turning point at (0, −3) and a point of inflexion at (−4, 2).

   (*a*) Sketch the graph of $y = f(-x)$.

   (*b*) On the same diagram, sketch the graph of $y = 2f(-x)$.

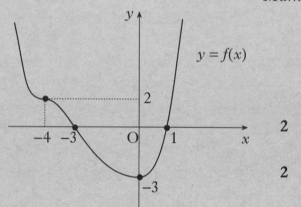

2

2

6. If $f(x) = \cos(2x) - 3\sin(4x)$, find the exact value of $f'\left(\dfrac{\pi}{6}\right)$.

4

7. Part of the graph of $y = 2\sin(x°) + 5\cos(x°)$ is shown in the diagram.

   (*a*) Express $y = 2\sin(x°) + 5\cos(x°)$ in the form $k\sin(x° + a°)$ where $k > 0$ and $0 \le a < 360$.

   (*b*) Find the coordinates of the minimum turning point P.

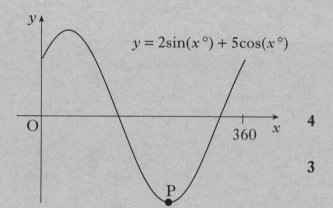

4

3

8. An open water tank, in the shape of a triangular prism, has a capacity of 108 litres. The tank is to be lined on the inside in order to make it watertight.

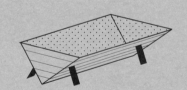

   The triangular cross-section of the tank is right-angled and isosceles, with equal sides of length *x* cm. The tank has a length of *l* cm.

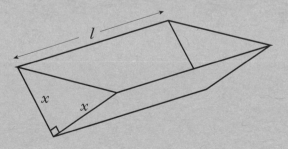

   (*a*) Show that the surface area to be lined, A cm^2, is given by $A(x) = x^2 + \dfrac{432000}{x}$.

3

   (*b*) Find the value of *x* which minimises this surface area.

5

Marks

9. The diagram shows vectors **a** and **b**.
   If $|\mathbf{a}| = 5$, $|\mathbf{b}| = 4$ and $\mathbf{a}.(\mathbf{a} + \mathbf{b}) = 36$, find the
   size of the acute angle θ between **a** and **b**.

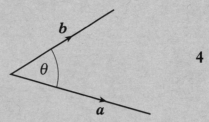

4

10. Solve the equation $3\cos(2x) + 10\cos(x) - 1 = 0$ for $0 \le x \le \pi$, correct to 2 decimal
    places.

5

11. (*a*) (i) Sketch the graph of $y = a^x + 1$, $a > 2$.

    (ii) On the same diagram, sketch the graph of $y = a^{x+1}$, $a > 2$.

2

   (*b*) Prove that the graphs intersect at a point where the x-coordinate is
   $\log_a\left(\dfrac{1}{a-1}\right)$.

3

[*END OF QUESTION PAPER*]

[BLANK PAGE]

2004 | Higher

[BLANK PAGE]

X100/301

NATIONAL
QUALIFICATIONS
2004

FRIDAY, 21 MAY
9.00 AM – 10.10 AM

MATHEMATICS
HIGHER
Units 1, 2 and 3
Paper 1
(Non-calculator)

Read Carefully

1 **Calculators may <u>NOT</u> be used in this paper.**

2 Full credit will be given only where the solution contains appropriate working.

3 Answers obtained by readings from scale drawings will not receive any credit.

SCOTTISH
QUALIFICATIONS
AUTHORITY

FORMULAE LIST

Circle:

The equation $x^2 + y^2 + 2gx + 2fy + c = 0$ represents a circle centre $(-g, -f)$ and radius $\sqrt{g^2 + f^2 - c}$.

The equation $(x - a)^2 + (y - b)^2 = r^2$ represents a circle centre (a, b) and radius r.

Scalar Product: $\quad a.b = |a|\,|b|\cos\theta$, where θ is the angle between a and b

$$\text{or}\quad a.b = a_1 b_1 + a_2 b_2 + a_3 b_3 \text{ where } a = \begin{pmatrix} a_1 \\ a_2 \\ a_3 \end{pmatrix} \text{ and } b = \begin{pmatrix} b_1 \\ b_2 \\ b_3 \end{pmatrix}.$$

Trigonometric formulae:

$$\sin(A \pm B) = \sin A \cos B \pm \cos A \sin B$$
$$\cos(A \pm B) = \cos A \cos B \mp \sin A \sin B$$
$$\sin 2A = 2\sin A \cos A$$
$$\cos 2A = \cos^2 A - \sin^2 A$$
$$= 2\cos^2 A - 1$$
$$= 1 - 2\sin^2 A$$

Table of standard derivatives:

$f(x)$	$f'(x)$
$\sin ax$	$a\cos ax$
$\cos ax$	$-a\sin ax$

Table of standard integrals:

$f(x)$	$\int f(x)\,dx$
$\sin ax$	$-\dfrac{1}{a}\cos ax + C$
$\cos ax$	$\dfrac{1}{a}\sin ax + C$

ALL questions should be attempted.

Marks

1. The point A has coordinates $(7, 4)$. The straight lines with equations $x + 3y + 1 = 0$ and $2x + 5y = 0$ intersect at B.

   (a) Find the gradient of AB. — **3**

   (b) Hence show that AB is perpendicular to only one of these two lines. — **5**

2. $f(x) = x^3 - x^2 - 5x - 3$.

   (a) (i) Show that $(x + 1)$ is a factor of $f(x)$.

   (ii) Hence or otherwise factorise $f(x)$ fully. — **5**

   (b) One of the turning points of the graph of $y = f(x)$ lies on the x-axis. Write down the coordinates of this turning point. — **1**

3. Find all the values of x in the interval $0 \le x \le 2\pi$ for which $\tan^2(x) = 3$. — **4**

4. The diagram shows the graph of $y = g(x)$.

   (a) Sketch the graph of $y = -g(x)$. — **2**

   (b) On the same diagram, sketch the graph of $y = 3 - g(x)$. — **2**

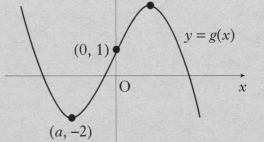

5. A, B and C have coordinates $(-3, 4, 7)$, $(-1, 8, 3)$ and $(0, 10, 1)$ respectively.

   (a) Show that A, B and C are collinear. — **3**

   (b) Find the coordinates of D such that $\overrightarrow{AD} = 4\overrightarrow{AB}$. — **2**

6. Given that $y = 3\sin(x) + \cos(2x)$, find $\dfrac{dy}{dx}$. — **3**

[Turn over for Questions 7 to 11 on *Page four*

Marks

7. Find $\int_0^2 \sqrt{4x+1}\ dx$.   **5**

8. (a) Write $x^2 - 10x + 27$ in the form $(x+b)^2 + c$.   **2**

   (b) Hence show that the function $g(x) = \frac{1}{3}x^3 - 5x^2 + 27x - 2$ is always increasing.   **4**

9. Solve the equation $\log_2(x+1) - 2\log_2(3) = 3$.   **4**

10. In the diagram
angle DEC = angle CEB = $x°$ and
angle CDE = angle BEA = $90°$.
CD = 1 unit; DE = 3 units.

By writing angle DEA in terms of $x°$, find the exact value of $\cos(\hat{DEA})$.   **7**

11. The diagram shows a parabola passing through the points $(0, 0)$, $(1, -6)$ and $(2, 0)$.

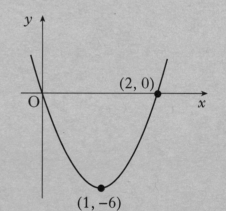

   (a) The equation of the parabola is of the form $y = ax(x-b)$.

      Find the values of a and b.   **3**

   (b) This parabola is the graph of $y = f'(x)$.

      Given that $f(1) = 4$, find the formula for $f(x)$.   **5**

[END OF QUESTION PAPER]

X100/303

NATIONAL
QUALIFICATIONS
2004

FRIDAY, 21 MAY
10.30 AM – 12.00 NOON

MATHEMATICS
HIGHER
Units 1, 2 and 3
Paper 2

Read Carefully

1 **Calculators may be used in this paper.**

2 Full credit will be given only where the solution contains appropriate working.

3 Answers obtained by readings from scale drawings will not receive any credit.

SCOTTISH
QUALIFICATIONS
AUTHORITY

FORMULAE LIST

Circle:

The equation $x^2 + y^2 + 2gx + 2fy + c = 0$ represents a circle centre $(-g, -f)$ and radius $\sqrt{g^2 + f^2 - c}$.

The equation $(x - a)^2 + (y - b)^2 = r^2$ represents a circle centre (a, b) and radius r.

Scalar Product: $\quad a.b = |a|\,|b| \cos\theta$, where θ is the angle between a and b

or $\quad a.b = a_1b_1 + a_2b_2 + a_3b_3$ where $a = \begin{pmatrix} a_1 \\ a_2 \\ a_3 \end{pmatrix}$ and $b = \begin{pmatrix} b_1 \\ b_2 \\ b_3 \end{pmatrix}$.

Trigonometric formulae:

$$\sin(A \pm B) = \sin A \cos B \pm \cos A \sin B$$
$$\cos(A \pm B) = \cos A \cos B \mp \sin A \sin B$$
$$\sin 2A = 2\sin A \cos A$$
$$\cos 2A = \cos^2 A - \sin^2 A$$
$$= 2\cos^2 A - 1$$
$$= 1 - 2\sin^2 A$$

Table of standard derivatives:

$f(x)$	$f'(x)$
$\sin ax$	$a\cos ax$
$\cos ax$	$-a\sin ax$

Table of standard integrals:

$f(x)$	$\int f(x)dx$
$\sin ax$	$-\dfrac{1}{a}\cos ax + C$
$\cos ax$	$\dfrac{1}{a}\sin ax + C$

ALL questions should be attempted.

Marks

1. (a) The diagram shows line OA with equation $x - 2y = 0$.

   The angle between OA and the x-axis is $a°$.

   Find the value of a.

   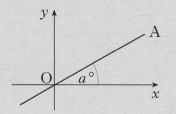

   3

   (b) The second diagram shows lines OA and OB. The angle between these two lines is $30°$.

   Calculate the gradient of line OB correct to 1 decimal place.

   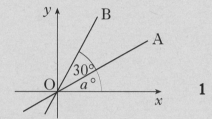

   1

2. P, Q and R have coordinates $(1, 3, -1)$, $(2, 0, 1)$ and $(-3, 1, 2)$ respectively.

   (a) Express the vectors $\overrightarrow{QP}$ and $\overrightarrow{QR}$ in component form. 2

   (b) Hence or otherwise find the size of angle PQR. 5

3. Prove that the roots of the equation $2x^2 + px - 3 = 0$ are real for all values of p. 4

4. A sequence is defined by the recurrence relation $u_{n+1} = ku_n + 3$.

   (a) Write down the condition on k for this sequence to have a limit. 1

   (b) The sequence tends to a limit of 5 as $n \to \infty$. Determine the value of k. 3

5. The point P(x, y) lies on the curve with equation $y = 6x^2 - x^3$.

   (a) Find the value of x for which the gradient of the tangent at P is 12. 5

   (b) Hence find the equation of the tangent at P. 2

[Turn over

Marks

6. (a) Express $3\cos(x°) + 5\sin(x°)$ in the form $k\cos(x° - a°)$ where $k > 0$ and $0 \le a \le 90$.   **4**

   (b) Hence solve the equation $3\cos(x°) + 5\sin(x°) = 4$ for $0 \le x \le 90$.   **3**

7. The graph of the cubic function $y = f(x)$ is shown in the diagram. There are turning points at $(1, 1)$ and $(3, 5)$.

   Sketch the graph of $y = f'(x)$.   **3**

   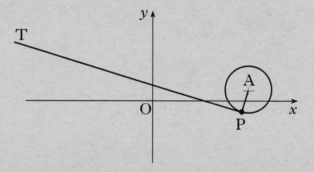

8. The circle with centre A has equation $x^2 + y^2 - 12x - 2y + 32 = 0$. The line PT is a tangent to this circle at the point $P(5, -1)$.

   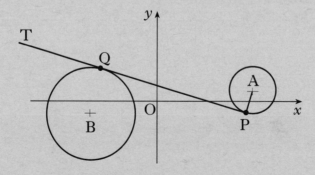

   (a) Show that the equation of this tangent is $x + 2y = 3$.   **4**

   The circle with centre B has equation $x^2 + y^2 + 10x + 2y + 6 = 0$.

   (b) Show that PT is also a tangent to this circle.   **5**

   (c) Q is the point of contact. Find the length of PQ.   **2**

Marks

9. An open cuboid measures internally
x units by $2x$ units by h units and has
an inner surface area of 12 units^2.

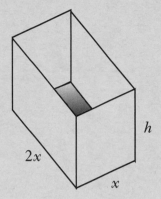

(a) Show that the volume, V units^3, of the cuboid is given by $V(x) = \frac{2}{3}x(6 - x^2)$.   **3**

(b) Find the exact value of x for which this volume is a maximum.   **5**

10. The amount A_t micrograms of a certain radioactive substance remaining after t years decreases according to the formula $A_t = A_0 e^{-0.002t}$, where A_0 is the amount present initially.

(a) If 600 micrograms are left after 1000 years, how many micrograms were present initially?   **3**

(b) The half-life of a substance is the time taken for the amount to decrease to half of its initial amount. What is the half-life of this substance?   **4**

11. An architectural feature of a
building is a wall with arched
windows. The curved edge of
each window is parabolic.

The second diagram shows
one such window. The shaded
part represents the glass.

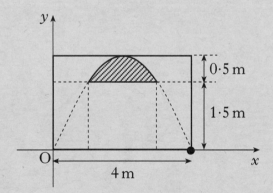

The top edge of the window is
part of the parabola with
equation $y = 2x - \frac{1}{2}x^2$.

Find the area in square metres
of the glass in one window.   **8**

[END OF QUESTION PAPER]

[BLANK PAGE]

2005 | Higher

[BLANK PAGE]

X100/301

NATIONAL
QUALIFICATIONS
2005

FRIDAY, 20 MAY
9.00 AM – 10.10 AM

MATHEMATICS
HIGHER
Units 1, 2 and 3
Paper 1
(Non-calculator)

Read Carefully

1 **Calculators may <u>NOT</u> be used in this paper.**

2 Full credit will be given only where the solution contains appropriate working.

3 Answers obtained by readings from scale drawings will not receive any credit.

SCOTTISH
QUALIFICATIONS
AUTHORITY

FORMULAE LIST

Circle:

The equation $x^2 + y^2 + 2gx + 2fy + c = 0$ represents a circle centre $(-g, -f)$ and radius $\sqrt{g^2 + f^2 - c}$.

The equation $(x - a)^2 + (y - b)^2 = r^2$ represents a circle centre (a, b) and radius r.

Scalar Product: $\mathbf{a}.\mathbf{b} = |\mathbf{a}|\,|\mathbf{b}|\cos\theta$, where θ is the angle between $\mathbf{a}$ and $\mathbf{b}$

or $\mathbf{a}.\mathbf{b} = a_1b_1 + a_2b_2 + a_3b_3$ where $\mathbf{a} = \begin{pmatrix} a_1 \\ a_2 \\ a_3 \end{pmatrix}$ and $\mathbf{b} = \begin{pmatrix} b_1 \\ b_2 \\ b_3 \end{pmatrix}$.

Trigonometric formulae:

$$\sin(A \pm B) = \sin A \cos B \pm \cos A \sin B$$
$$\cos(A \pm B) = \cos A \cos B \mp \sin A \sin B$$
$$\sin 2A = 2\sin A \cos A$$
$$\cos 2A = \cos^2 A - \sin^2 A$$
$$= 2\cos^2 A - 1$$
$$= 1 - 2\sin^2 A$$

Table of standard derivatives:

$f(x)$	$f'(x)$
$\sin ax$	$a\cos ax$
$\cos ax$	$-a\sin ax$

Table of standard integrals:

$f(x)$	$\int f(x)dx$
$\sin ax$	$-\frac{1}{a}\cos ax + C$
$\cos ax$	$\frac{1}{a}\sin ax + C$

ALL questions should be attempted.

Marks

1. Find the equation of the line ST, where T is the point (–2, 0) and angle STO is 60°.

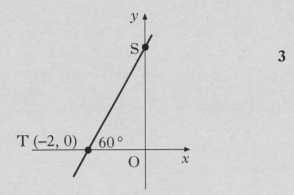

3

2. Two congruent circles, with centres A and B, touch at P.

   Relative to suitable axes, their equations are

   $x^2 + y^2 + 6x + 4y - 12 = 0$ and
   $x^2 + y^2 - 6x - 12y + 20 = 0$.

   (a) Find the coordinates of P.

   (b) Find the length of AB.

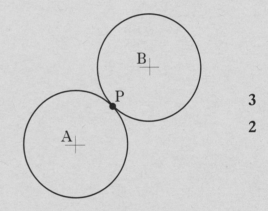

3

2

3. D,OABC is a pyramid. A is the point (12, 0, 0), B is (12, 6, 0) and D is (6, 3, 9).

   F divides DB in the ratio 2:1.

   (a) Find the coordinates of the point F.

   (b) Express $\overrightarrow{AF}$ in component form.

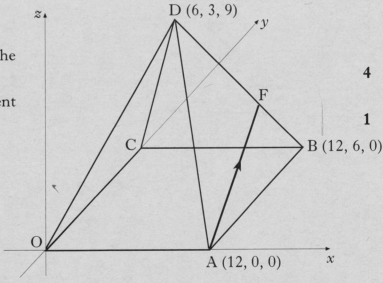

4

1

[Turn over

Marks

4. Functions $f(x) = 3x - 1$ and $g(x) = x^2 + 7$ are defined on the set of real numbers.

   (a) Find $h(x)$ where $h(x) = g(f(x))$.   2

   (b)   (i)   Write down the coordinates of the minimum turning point of $y = h(x)$.

        (ii)   Hence state the range of the function h.   2

5. Differentiate $(1 + 2\sin x)^4$ with respect to x.   2

6. (a) The terms of a sequence satisfy $u_{n+1} = ku_n + 5$. Find the value of k which produces a sequence with a limit of 4.   2

   (b) A sequence satisfies the recurrence relation $u_{n+1} = mu_n + 5$, $u_0 = 3$.

        (i)   Express u_1 and u_2 in terms of m.

        (ii)   Given that $u_2 = 7$, find the value of m which produces a sequence with no limit.   5

7. The function f is of the form $f(x) = \log_b(x - a)$. The graph of $y = f(x)$ is shown in the diagram.

   (a) Write down the values of a and b.   2

   (b) State the domain of f.   1

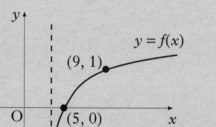

8. A function f is defined by the formula $f(x) = 2x^3 - 7x^2 + 9$ where x is a real number.

   (a) Show that $(x - 3)$ is a factor of $f(x)$, and hence factorise $f(x)$ fully.   5

   (b) Find the coordinates of the points where the curve with equation $y = f(x)$ crosses the x- and y-axes.   2

   (c) Find the greatest and least values of f in the interval $-2 \le x \le 2$.   5

9. If $\cos 2x = \dfrac{7}{25}$ and $0 < x < \dfrac{\pi}{2}$, find the exact values of $\cos x$ and $\sin x$.   4

Marks

10. (a) Express $\sin x - \sqrt{3}\cos x$ in the form $k\sin(x-a)$ where $k > 0$ and $0 \le a \le 2\pi$.   **4**

     (b) Hence, or otherwise, sketch the curve with equation $y = 3 + \sin x - \sqrt{3}\cos x$ in the interval $0 \le x \le 2\pi$.   **5**

11. (a) A circle has centre $(t, 0)$, $t > 0$, and radius 2 units.

        Write down the equation of the circle.   **1**

     (b) Find the exact value of t such that the line $y = 2x$ is a tangent to the circle.   **5**

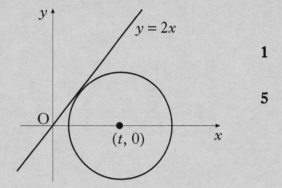

[END OF QUESTION PAPER]

[BLANK PAGE]

X100/303

NATIONAL
QUALIFICATIONS
2005

FRIDAY, 20 MAY
10.30 AM – 12.00 NOON

MATHEMATICS
HIGHER
Units 1, 2 and 3
Paper 2

Read Carefully

1   **Calculators may be used in this paper.**

2   Full credit will be given only where the solution contains appropriate working.

3   Answers obtained by readings from scale drawings will not receive any credit.

SCOTTISH
QUALIFICATIONS
AUTHORITY

FORMULAE LIST

Circle:

The equation $x^2 + y^2 + 2gx + 2fy + c = 0$ represents a circle centre $(-g, -f)$ and radius $\sqrt{g^2 + f^2 - c}$.

The equation $(x - a)^2 + (y - b)^2 = r^2$ represents a circle centre (a, b) and radius r.

Scalar Product: $\quad a.b = |a|\,|b|\cos\theta$, where θ is the angle between a and b

or $\quad a.b = a_1b_1 + a_2b_2 + a_3b_3$ where $a = \begin{pmatrix} a_1 \\ a_2 \\ a_3 \end{pmatrix}$ and $b = \begin{pmatrix} b_1 \\ b_2 \\ b_3 \end{pmatrix}$.

Trigonometric formulae:

$$\sin(A \pm B) = \sin A \cos B \pm \cos A \sin B$$
$$\cos(A \pm B) = \cos A \cos B \mp \sin A \sin B$$
$$\sin 2A = 2\sin A \cos A$$
$$\cos 2A = \cos^2 A - \sin^2 A$$
$$= 2\cos^2 A - 1$$
$$= 1 - 2\sin^2 A$$

Table of standard derivatives:

$f(x)$	$f'(x)$
$\sin ax$	$a \cos ax$
$\cos ax$	$-a \sin ax$

Table of standard integrals:

$f(x)$	$\int f(x)dx$
$\sin ax$	$-\dfrac{1}{a}\cos ax + C$
$\cos ax$	$\dfrac{1}{a}\sin ax + C$

ALL questions should be attempted.

Marks

1. Find $\int \dfrac{4x^3 - 1}{x^2}\, dx, \quad x \neq 0.$ 4

2. Triangles ACD and BCD are right-angled at D with angles p and q and lengths as shown in the diagram.

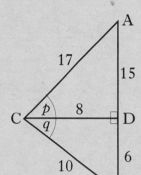

   (a) Show that the exact value of $\sin(p + q)$ is $\dfrac{84}{85}$.   4

   (b) Calculate the exact values of:

        (i) $\cos(p + q)$;

        (ii) $\tan(p + q)$.   3

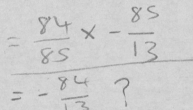

3. (a) A chord joins the points A(1,0) and B(5,4) on the circle as shown in the diagram.

   Show that the equation of the perpendicular bisector of chord AB is $x + y = 5$.   4

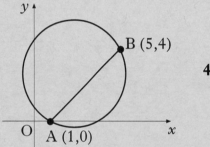

   (b) The point C is the centre of this circle. The tangent at the point A on the circle has equation $x + 3y = 1$.

   Find the equation of the radius CA.   4

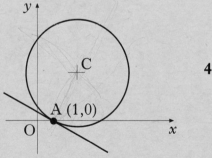

   (c)   (i) Determine the coordinates of the point C.

        (ii) Find the equation of the circle.   4

[Turn over

Marks

4. The sketch shows the positions of Andrew(A), Bob(B) and Tracy(T) on three hill-tops.

   Relative to a suitable origin, the coordinates (in hundreds of metres) of the three people are A(23, 0, 8), B(–12, 0, 9) and T(28, –15, 7).

   In the dark, Andrew and Bob locate Tracy using heat-seeking beams.

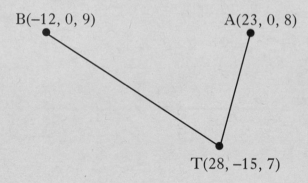

B(–12, 0, 9)   A(23, 0, 8)

T(28, –15, 7)

   (a) Express the vectors $\overrightarrow{TA}$ and $\overrightarrow{TB}$ in component form.   2

   (b) Calculate the angle between these two beams.   5

5. The curves with equations $y = x^2$ and $y = 2x^2 - 9$ intersect at K and L as shown.

   Calculate the area enclosed between the curves.   8

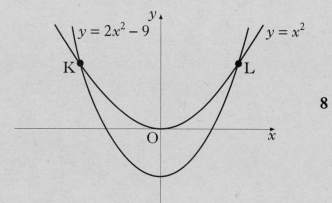

6. The diagram shows the graph of $y = \dfrac{24}{\sqrt{x}}$, $x > 0$.

   Find the equation of the tangent at P, where $x = 4$.   6

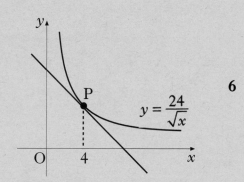

7. Solve the equation $\log_4(5 - x) - \log_4(3 - x) = 2$, $x < 3$.   4

Marks

8. Two functions, f and g, are defined by $f(x) = k\sin 2x$ and $g(x) = \sin x$ where $k > 1$.

The diagram shows the graphs of $y = f(x)$ and $y = g(x)$ intersecting at O, A, B, C and D.

Show that, at A and C, $\cos x = \dfrac{1}{2k}$.

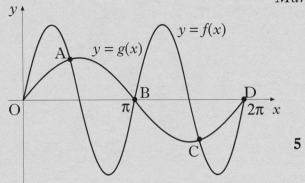

5

9. The value V (in £ million) of a cruise ship t years after launch is given by the formula $V = 252e^{-0\cdot06335t}$.

   (*a*) What was its value when launched?   1

   (*b*) The owners decide to sell the ship once its value falls below £20 million. After how many years will it be sold?   4

10. Vectors a and c are represented by two sides of an equilateral triangle with sides of length 3 units, as shown in the diagram.

Vector b is 2 units long and b is perpendicular to both a and c.

Evaluate the scalar product $a.(a + b + c)$.   4

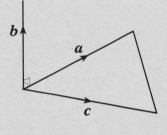

11. (*a*) Show that $x = -1$ is a solution of the cubic equation $x^3 + px^2 + px + 1 = 0$.   1

   (*b*) Hence find the range of values of p for which all the roots of the cubic equation are real.   7

[*END OF QUESTION PAPER*]

[BLANK PAGE]

[BLANK PAGE]

[BLANK PAGE]

[BLANK PAGE]

[BLANK PAGE]

[BLANK PAGE]

[BLANK PAGE]